Images of Modern America

JOINT BASE LANGLEY-EUSTIS

The Joint Base Langley-Eustis (JBLE) Mission Patch. The Joint Base Langley-Eustis mission patch shows a pair of Langley Air Force Base (LAFB)–based Lockheed Martin F-22A Raptors in flight and a US Army soldier from Fort Eustis, in nearby Newport News, Virginia. As of 2013, JBLE employed 17,162 military workers, 6,355 civil servants, and 27,256 contractors. (Courtesy of the US Air Force [USAF].)

Front Cover: Ground crews perform maintenance on a US Army Air Forces Douglas A-20 Havoc attack bomber at Langley Field, Virginia, in July 1942. (Courtesy of the US Library of Congress, Alfred T. Palmer.)

Upper Back Cover: 47th Light Bomb Wing B-45A-5-NA Tornadoes on the flight line at Langley Air Force Base in 1952. (Courtesy of the USAF.)

Lower Back Cover (from left to right): the *Roma* airship rises above Langley Field on her ill-fated flight of February 21, 1922 (courtesy of the USAF, Air Combat Command History Office); a couple of USAF Lockheed Martin F-22 Raptors fly over JBLE on July 24, 2013 (courtesy of the USAF, S.Sgt. John D. Strong II); the "Mercury Seven," America's first seven astronauts who took part in Project Mercury, pose for a publicity photograph in front of a LAFB-based Convair F-106B Delta Dart supersonic jet interceptor on January 20, 1961. (Courtesy of NASA.)

JOINT BASE LANGLEY-EUSTIS

Mark A. Chambers

Images of Modern America

ARCADIA PUBLISHING

ISBN 978-1-4671-2543-7

Published by Arcadia Publishing
Charleston, South Carolina

Printed in the United States of America

Library of Congress Control Number: 2016953989

For all general information, please contact Arcadia Publishing:
Telephone 843-853-2070
Fax 843-853-0044
E-mail sales@arcadiapublishing.com
For customer service and orders:
Toll-Free 1-888-313-2665

Visit us on the Internet at www.arcadiapublishing.com

This pictorial history is dedicated to the pioneering spirit and courage of the servicemen and -women of Joint Base Langley-Eustis who have helped to shape and mold American military aviation for over a century.

Contents

ACKNOWLEDGMENTS

I would like to thank several important individuals who contributed to this fascinating pictorial history of one of our nation's oldest Air Force bases. I would like to thank my wife, Lesa; daughter, Caitlyn; and sons, Patrick and Ryan, for their constant support and patience with me during the preparation of this book. I would also like to thank William Butler, deputy command historian, Headquarters Air Combat Command, Joint Base Langley-Eustis, for providing invaluable photographic support as well as an electronic copy of JBLE's centennial celebration publication, *Langley: A Centennial History, 1916–2016*.

As always, many thanks go to Holly Reed and the staff of the US National Archives at College Park, Maryland, Still Pictures Branch, for photographic support for this project.

Special thanks also go to Mary Scott Hardaway, acquisitions editor, Arcadia Publishing, for securing publication of this work and to Liz Gurley, senior title manager, Arcadia Publishing, for her invaluable editorial assistance and fantastic support for this project.

Introduction

For over a century, Joint Base Langley-Eustis (JBLE) has served as a cornerstone for the advancement of Air Force aviation in America. This historic Air Force installation has helped keep America on the cutting edge of military aviation while proving to be vital to national defense and security. From helping to facilitate the growth and influence of Army airpower during the 1920s, helping to win World War II, and halting the spread of communism during the Korean War to quelling Cold War tensions during the Cuban Missile Crisis, supporting the American effort during the Vietnam War, and ejecting Saddam Hussein's Iraqi military forces from Kuwait during Operation Desert Storm, JBLE has helped to extend the awesome impact of American airpower throughout the world. JBLE continued to make its presence felt in the War on Terror as well as in Operation Iraqi Freedom and will serve as an essential player in shaping America's Air Force of the future.

On May 6, 1896, the brilliant professor and inventor Samuel Pierpont Langley, third secretary of the Smithsonian Institution, conducted a successful series of flight tests of his Aerodrome No. 5 powered flying model airplane, catapult launched from a houseboat on the peaceful waters of the Potomac River close to Quantico, Virginia. During the unprecedented test flights, the aircraft achieved altitudes of 3,300 feet and 2,300 feet on consecutive flights. The experiments marked the first time that an engine-powered heavier-than-air craft successfully achieved significant altitudes on consecutive flights. Langley then became more emboldened and ambitious, and on October 7, 1903, attempted to duplicate the feats achieved by Aerodrome No. 5, but with a full-scale, manned version, once again to be launched via catapult from a houseboat on the Potomac. Two unsuccessful attempts, with Charles Manly at the controls, were made from a massive catapult track aboard the houseboat. Manly managed to survive both failed flight attempts. Despite these failures of the full-scale, manned Aerodrome, Langley became widely recognized throughout the world as an American aviation pioneer. He was so highly respected and revered that the US Army Signal Corps paid him the ultimate honor by naming one of its first experimental flying fields, in Hampton, Virginia, Langley Field.

In the summer of 1916, the National Advisory Committee for Aeronautics (NACA)—the precursor agency to today's National Aeronautics and Space Administration (NASA), which also included members of the Smithsonian Institution, the Army Signal Corps Aviation Section, other government departments, and aviation industry leaders—proposed the establishment of a unique experimental airfield and aviation research laboratory that was to be utilized by the NACA, Army, and Navy. On December 30, the US government purchased 2,500 acres of land upon which numerous plantations once existed, including the famous Wythe plantation. The Army Signal Corps Aviation Section ultimately purchased the land at a cost of $300,000 under the Army Appropriation Act of 1917. The total amount of land designated to comprise the new airfield and laboratory was a little over 1,600 acres, located near the Back River in Hampton. In June 1917, military personnel arrived at the site of the new airfield and research laboratory. The first organization to establish operations at the new air base was the 5th Aviation School, Signal Corps. This unit was re-designated the 119th Aero Squadron in September, and in November, the US Army School of Aerial Photography and a Photo School detachment were established at the field. The air base was officially designated Langley Field on August 7, 1917. During World War I, in addition to determining the best Allied aircraft procurement prospects for the US Army Signal Corps, Langley Field also served as a flight test center for US Navy seaplanes and observation/spotter aircraft.

Among the seaplanes based at Langley was a small unit comprised of Italian-made Macchi flying boats that were piloted by some pioneers in US naval aviation. Among these pioneers was Lt.

Charles Hazeltine Hammann (US Naval Reserve Force), who became the first US naval aviator to be awarded the Medal of Honor for uncommon valor demonstrated in combat in Europe during World War I. Tragically, Lieutenant Hammann was killed when his Macchi M.5 crashed at Langley Field on June 24, 1919. A commemorative plaque marks the exact crash site at JBLE today.

On March 3, 1919, the first balloon units, the US Army Air Service 61st and 71st Balloon Companies, arrived at Langley Field to conduct aerial observation operations. All the balloons and eventually the much larger airships that arrived at Langley Field used hydrogen gas as a lifting medium, which came from a gas generating and compressor plant erected in 1917. In 1919, construction work commenced on an airship hangar, as well as on the base headquarters building, which would later serve as the home offices of the USAF Tactical Air Command (TAC) and eventually Air Combat Command (ACC).

In 1921, the US Army and Navy, with authorization and approval of the War Department, jointly decided to conduct experimental flight tests to evaluate the effectiveness of airplanes when pitted against battleships in combat. The US Army Air Service's Brig. Gen. William "Billy" Mitchell, an outspoken airpower advocate, was invited by the US Navy to participate in a series of battleship bombing flight experiments to be conducted in the coastal waters of Virginia. The Air Service Supply Detachment, 1st and 5th Aero Observation Squadrons, originally based at Mitchell Field, New York, as well as the 49th and 96th Bombardment Squadrons, were detached to Langley Field to participate in the experiments. At the same time, Brigadier General Mitchell formed the 1st Provisional Air Brigade at Langley Field specifically to take part in the battleship bombings. The historic battleship bombing experiments were initiated on July 18, with the successful bombing and sinking of the ex-German cruiser *Frankfurt*, which was at anchor about 100 miles out in the Atlantic Ocean from the entrance to the Chesapeake Bay. On July 21, the ex-German battleship *Ostfriesland* was also bombed and sent to the bottom of the Atlantic. This sinking effectively proved the US Navy and Army's theory that the airplane was superior to warships and that warships were vulnerable to air attack.

On October 1, 1921, the 10th and 19th Airship Companies, which now included older balloon companies, were consolidated and comprised the first Airship School at Langley Field. Another lighter-than-air (LTA) milestone was achieved on November 15, 1921, when the large, Italian-made *Roma* airship was test flown for the first time at Langley. The mammoth airship was the largest to be operated by the US military and was of the semi-rigid type. It utilized hydrogen as a lifting medium. Tragedy struck, however, on February 21, 1922, when the *Roma*, carrying 45 crewmembers on a routine operational flight over Norfolk, Virginia, experienced technical difficulties in-flight (the large box-kite rudder slipped askew), and the airship lost altitude and struck power lines, igniting the flammable hydrogen within the airship's envelope, which in turn caused a huge explosion. A total of 34 crewmembers perished in what was the worst air disaster in US history at the time. Following this tragedy, the US military only used helium for its airships.

By 1923, Langley Field's facilities had grown. There were 26 officer housing units, two aircraft hangars, a massive airship hangar, two BOQs (bachelor officer's quarters), a seaplane hangar, and a boathouse. Another significant milestone in aviation history was achieved that year, when Brig. Gen. "Billy" Mitchell authorized a precedent-setting around-the-world flight, using four specially designed Douglas aircraft to accomplish the feat. Aircrews commenced training at Langley Field for the historic mission in December 1923. The crews manning the four aircraft, named the *Boston*, *Seattle*, *Chicago*, and *New Orleans*, launched from Seattle, Washington, on April 6, 1924, successfully returning to Seattle on September 28, 1924. In September 1923, another milestone was achieved in aviation history when precedent-setting parasite flight experiments were successfully performed by airships and biplane aircraft (the "parasites") in the skies over Langley Field, effectively demonstrating the feasibility of the "aerial aircraft carrier" concept.

During the 1930s, notable Army Air Corps aircraft such as the Martin B-10 bomber, Curtiss P-36 Hawk, the Boeing YB-17 Flying Fortress bomber prototype, and the massive experimental Boeing XB-15 bomber operated at Langley Field. The Flying Fortress prototype was detached to Langley as part of the 2nd Bombardment Group on March 1, 1937, to help train bomber crews so that they could become familiar with B-17 flight operations and tactical training. From February 14 to 15, 1939, the Boeing XB-15 bomber performed another aviation history first when it successfully transported 3,250 pounds of medical supplies from Langley Field to Chile to relieve the country's earthquake-ravaged populace. The historic airlift mission was performed nonstop and lasted over 29 hours.

The 1st Sea Search Attack Group was established at Langley Field on June 8, 1942, to pioneer revolutionary new aerial anti-submarine warfare (ASW) technologies. Flying modified Douglas B-18B Bolo bombers, the ASW air unit pioneered the magnetic anomaly detector (MAD) boom. The MAD boom was effectively outfitted on both US Army Air Forces (USAAF) and US Navy patrol aircraft, resulting in the sinking of numerous German U-boats in both the Atlantic Ocean and European waters. Pioneering radar bombing studies were also conducted at Langley Field during the latter portion of World War II.

On May 25, 1946, the TAC Headquarters was created at Langley Field. TAC was established to organize, train, outfit, and maintain combat air units that could be deployed rapidly on short notice in times of both peace and war. In January 1948, Langley Field was renamed Langley Air Force Base (LAFB) following the creation of the US Air Force (USAF) in 1947. As of April 1949, the Continental Air Command at Langley comprised the 4th Fighter Wing, which flew Lockheed F-80 Shooting Stars and North American F-86A Sabres.

With the outbreak of hostilities in Korea on June 25, 1950, LAFB played a critical role, supporting US and United Nations (UN) forces against the communist forces of North Korea, China, and the Soviet Union by providing USAF fighter pilots with tactical training in the service's premier jet fighters. During the Korean War, the USAF's premier air superiority fighter and superb jet dogfighter, the North American F-86 Sabre, was stationed at LAFB.

During the frightening Cuban Missile Crisis, from October 16 to 28, 1962, USAF general Curtis LeMay pushed for the carpet bombing of Cuba should tensions escalate. As a result, USAF Strategic Air Command (SAC) Boeing B-47A Stratojet bombers, which had nuclear weapons–carrying capability, were detached to LAFB and were instructed to prepare to conduct bombings of the Soviet satellite state. Fortunately, the crisis was defused when the Soviet Union agreed to back down and removed its nuclear missiles from Cuba.

From 1963 to 1975, LAFB hosted several Lockheed C-130 Hercules military transport aircraft units. These units included the 463rd Troop Carrier Wing (1963), 316th Tactical Airlift Wing (1966), and the Military Airlift Command's (MAC) 316th Tactical Airlift Wing (1975). All the Hercules units stationed at LAFB supported America's involvement in the Vietnam War.

To meet global air superiority needs, the 1st Tactical Fighter Wing, operational at MacDill Air Force Base, Florida, was moved to LAFB in January 1976. The unit was also the first USAF unit to receive the new McDonnell Douglas F-15A Eagle air superiority fighter.

The 1980s and the "Peace through Strength" policy approach to national defense embraced by Pres. Ronald Reagan and his administration led to a great military buildup both nationally and at LAFB. Major base renovations and improvements commenced in February 1980 with the opening of a $4-million base commissary. An enhanced gatehouse and visitor center opened in 1982. In December 1985, a state-of-the-art flight operations control tower became operational. In 1987, a unique, remote LAFB primary care clinic came online at nearby Bethel Manor.

The Reagan years also gave a much-needed boost to national defense operations at LAFB. In 1981, the 48th Fighter Interceptor Squadron (FIS) transitioned from the venerable F-106 to the advanced F-15, and in January 1982, a 1st Tactical Fighter Wing F-15 Eagle became the first aircraft of its type to intercept a Soviet Tu-95 "Bear" long-range bomber.

Iraq invaded Kuwait on August 2, 1990, and occupied that nation for seven months. Shortly following this act of aggression in the always volatile Middle East, the United States and its partner nations began amassing a formidable coalition military force designed to overwhelm the Iraqi dictator Saddam Hussein's forces and ultimately eject them from Kuwait. The Allied Coalition military buildup assembled in the Arabian deserts surrounding both Iraq and Kuwait in Operation Desert Shield. Later in August, a squadron of stealth fighters, yet to be battle-tested, was deployed from Tonopah Test Range, Nevada, to an air base in Saudi Arabia. On their way to the Middle East, the Nighthawks stopped at LAFB, refueling and receiving service. These same Nighthawks spearheaded the initial bombing missions against Iraq during Operation Desert Storm, which commenced on January 16, 1991. On August 7 and 8, 1990, the LAFB-based 71st and 27th Tactical Fighter Squadrons were deployed to Dhahran, Saudi Arabia, to serve as initial air superiority elements of Operation Desert Shield. Both squadrons conducted initial combat sorties on January 17, 1991.

Langley's host air wing was renamed the 1st Fighter Wing on October 1, 1991. On June 1, TAC Headquarters was inactivated and ACC established at LAFB. In December 1997, a $9-million, state-of-the-art civil engineering building was erected at LAFB.

On September 11, 2001, two LAFB-based Air National Guard F-16 Fighting Falcons took off from Langley to patrol the skies above Washington, DC, following the devastating hijacked airliner suicide attacks orchestrated by terrorists on the World Trade Center in New York City and the Pentagon. Unfortunately, the Fighting Falcons were unable to reach the air space above the Pentagon in time to prevent the tragedy that occurred there. Throughout the remainder of the first decade of the 2000s, LAFB-based F-15s participated in Operation Noble Eagle, which involved routine anti-air-based terrorist threat patrol sorties over Washington and the East Coast of the United States.

LAFB-based F-15s continued patrolling the no-fly zone in Iraq during the early 2000s and participated in Operation Iraqi Freedom in 2003, which led to the ultimate defeat of Iraq and the end of Saddam Hussein's brutal reign.

On May 12, 2005, the LAFB-based 1st Fighter Wing became the first USAF squadron to operate the new, stealthy Lockheed Martin F-22A Raptor air superiority fighter, and on December 15, 2005, the LAFB-based 27th Fighter Squadron achieved combat operational status with the F-22A Raptor.

On January 7, 2010, elements from the Army base at Fort Eustis and LAFB were combined to form Joint Base Langley-Eustis (JBLE). The newly formed 633rd Air Base Wing was designated to command joint base operations at LAFB.

From December 2 to 18, 2015, JBLE (1st Fighter Wing) served as the home base for a trilateral exercise, consisting of USAF, British Royal Air Force (RAF), and French Air Force aircraft and personnel elements. Aircraft utilized in the training exercise included the USAF F-22, RAF Eurofighter Typhoon, and French Air Force Dassault Rafale.

One

Growing Army Air Service 1896–1941

On May 6, 1896, Samuel Pierpont Langley, third secretary of the Smithsonian Institution, conducted successful flight tests of his Aerodrome No. 5 powered model airplane, catapult-launched from a houseboat on the Potomac River close to Quantico, Virginia. On October 7, 1903, Langley attempted to successfully flight test a full-scale, manned version of his Aerodrome, once again to be launched via catapult from a houseboat on the Potomac. Two test flights were attempted but failed. Langley was later honored by the US Army Signal Corps when it named one of its first experimental flying fields after him.

In June 1917, military personnel arrived at the new airfield and research laboratory, which was destined to become one of America's first military air bases. The 119th Aero Squadron was formed in September, and in November, the US Army School of Aerial Photography, as well as a Photo School detachment, was established at the field. The air base was officially designated Langley Field on August 7, 1917. On March 3, 1919, the US Army Air Service 61st and 71st Balloon Companies arrived at Langley Field to conduct aerial observation operations. In 1919, construction was initiated on an airship hangar and base headquarters building.

In 1921, Brig. Gen. William "Billy" Mitchell oversaw the US Army efforts in battleship bombing tests to be conducted in the coastal waters of Virginia. General Mitchell formed the 1st Provisional Air Brigade at Langley Field. On July 21, the ex-German battleship *Ostfriesland* was bombed and sent to the bottom. This effectively proved the theory that warships were vulnerable to air attack.

An LTA milestone was achieved on November 15, 1921, when the *Roma* airship was test flown for the first time at Langley. The huge airship used hydrogen as a lifting medium. It met its demise when it crashed and exploded in Norfolk on February 21, 1922.

During the 1930s, advanced Army Air Corps aircraft, including the Martin B-10 bomber, Curtiss P-36 Hawk, Boeing YB-17 Flying Fortress bomber prototype, and Boeing XB-15 bomber, were based at Langley Field.

JBLE FATHER AND AMERICAN AVIATION PIONEER. Samuel Pierpont Langley was the third secretary of the Smithsonian Institution. Having gained fame and respect for his successful Aerodrome No. 5 flights, the US Army Signal Corps decided to name its experimental flying field in Hampton in honor of him in 1916. (Courtesy of USAF, Air Combat Command History Office.)

THE LANGLEY AERODROME MODEL NO. 5. Professor Langley's flying model Aerodrome No. 5 is pictured in storage at a shop following its successful flights. (Courtesy of the US National Archives, Still Pictures Branch.)

Launched into History. On May 6, 1896, the steam-powered Aerodrome No. 5 was successfully launched from a specially designed spring catapult aboard a houseboat on the peaceful waters of the Potomac River, close to Quantico, Virginia. (Courtesy of the National Air and Space Museum [NASM], Smithsonian Institution, via NASA Langley Research Center.)

A Historic Flight. The Aerodrome No. 5 performs a successful flight on May 6, 1896. This photograph was taken by Alexander Graham Bell. This aeronautical feat has been recognized worldwide as the first flight of a heavier-than-air machine. (Courtesy of US National Archives, Still Pictures Branch.)

Prelude to a Flight-Test Failure. The full-scale, manned version of Langley's Aerodrome, seen here atop its houseboat launch platform, is prepped for launch on October 7, 1903. During this test flight attempt, the aircraft, with Charles Manly at the controls, was catapulted along a track aboard the houseboat but became ensnared in the launching device, causing the Aerodrome to pitch downward into the Potomac. (Courtesy of US National Archives, Still Pictures Branch.)

Second Flight-Test Failure. A second flight test attempt of Langley's full-scale, manned Aerodrome was made on December 8, 1903. This time, the aircraft lifted off the track nose up and experienced a catastrophic structural failure, plunging into the Potomac. The pilot, Charles Manly, survived. Despite the flight test failures of the manned Aerodrome, the US Army Signal Corps named its experimental flying field in Hampton Langley Field. (Courtesy of NASM, Smithsonian Institution, via NASA Langley Research Center.)

Langley Field Main Entrance. The Langley Field main entrance is pictured during World War I. Note the Thomas-Morse Scout biplane visible through the archway. (Courtesy of US National Archives, Still Pictures Branch.)

1918 Aerial Photography School Inspection. Photographic aircraft are lined up at an inspection of the Aerial Photography School at Langley Field in 1918. All of the aircraft are Curtiss JN-4H Jennies, with the exception of one Thomas-Morse Scout in the front row. (Courtesy of USAF, Air Combat Command History Office, via NASA Langley Research Center.)

US NAVY STANDARD H-4-H SEAPLANE. A US Navy Standard H-4-H seaplane is pictured at Langley Field during World War I. During the Great War, Langley Field was used by the US Navy as a base for experimental seaplanes. (Courtesy of USAF, Air Combat Command History Office, via NASA Langley Research Center.)

US NAVY ITALIAN MACCHI M.5 FLYING BOAT FIGHTER. A US Navy Macchi M.5 is at rest on the shore of the Back River at Langley Field in 1919. (Courtesy of Martin Copp, via NASA Langley Research Center.)

Lt. Charles Hazeltine Hammann's Macchi M.5 Flying Boat Fighter. Lieutenant Hammann's Italian Macchi M.5, armed with a pair of fixed machine guns, sits on the seaplane ramp at Langley Field in 1919. Lieutenant Hammann became the first US naval aviator to be awarded the Medal of Honor in Europe during World War I. Tragically, he was killed when his Macchi M.5 crashed during an air show while giving a demonstration simulating aerial combat at Langley Field on June 24, 1919. (Courtesy of Martin Copp, via NASA Langley Research Center.)

"Put 'em up!" In August 1919, the Knights of Columbus brought in ex-champion boxer Johnny Coulon to Langley Field to teach base airmen the finer points of the pugilistic profession. (Courtesy of US National Archives, Still Pictures Branch.)

"And in this Corner . . ." This boxing match was hosted by the Knights of Columbus at Langley Field on August 22, 1919. (Courtesy of US National Archives, Still Pictures Branch.)

Dirigible Hangar. Construction progresses on the dirigible hangar at Langley Field on April 23, 1920. (Courtesy of US National Archives, Still Pictures Branch.)

Aerial View of the Airship Hangar from an Airship. This view of the airship hangar at Langley Field was taken from the airship ZDUS-1 on September 3, 1920. The Z2 airship is visible inside the hangar. The airship hangar was torn down in 1947 and the site is now a parking lot. (Courtesy of US National Archives, Still Pictures Branch.)

Inside the Langley Field Dirigible Hangar, April 23, 1920. Note the US Navy C-2 airship, which made a transcontinental flight across America, as well as a kite balloon in the right foreground. (Courtesy of US National Archives, Still Pictures Branch.)

Rooftop View of the Langley Field Officers' Quarters. This view of the officers' quarters at Langley Field was taken in the spring of 1920. (Courtesy of US National Archives, Still Pictures Branch.)

Battleship Bombing Experiment Aircraft Line Up at Langley Field. Aircraft of the 1st Provisional Air Brigade form up on the flying field at Langley Field in July 1921, prior to taking part in the historic battleship bombing experiments conducted from July 18 to 21. (Courtesy of Martin Copp, via NASA Langley Research Center)

LEGENDARY US ARMY GENERALS. Gen. Billy Mitchell poses in front of the American flag beside Gen. John J. Pershing (right) for a publicity photograph taken in the review stands at Langley Field prior to the historic battleship bombing experiments in July 1921. (Courtesy of USAF, Air Combat Command History Office.)

BOMBED-UP BATTLESHIP-BOMBING BOMBER. A 1st Provisional Air Brigade Martin MB-2 bomber is loaded up with a 2,000-pound demolition bomb in preparation for the historic battleship bombing experiments in July 1921. (Courtesy of USAF, Air Combat Command History Office.)

A Direct Hit. A 2,000-pound bomb registers a direct hit on the Ex-German battleship *Ostfriesland* during the historic bombing experiments on July 21, 1921. The bomb was dropped by a 1st Provisional Air Brigade Martin MB-2 bomber. (Courtesy of Martin Copp, via NASA Langley Research Center.)

A 100-pound Bomb Strikes the USS Alabama. In September 1921, a second round of battleship bombing experiments was initiated, which resulted in the sinking of the World War I–vintage American battleship *Alabama* in the Chesapeake Bay close to Tangier Island on September 26, 1921. Here, a Langley Field–based Martin MB-2 bomber has dropped four 100-pound phosphorous bombs, which resulted in one direct hit on the dreadnought's crow's nest. (Courtesy of Martin Copp, via NASA Langley Research Center.)

Prepping the *Roma*. On November 15, 1921, the Italian-built *Roma* airship is "tether walked" out of the massive dirigible hangar at Langley Field for her first American test flight. (Courtesy of Martin Copp, via NASA Langley Research Center.)

The Langley LTA Hangar, 1922. Note the hydrogen/helium gas plant in the background to the right. (Courtesy of US National Archives, Still Pictures Branch.)

Prelude to an Air Disaster. This photograph was taken moments before the *Roma* crashed on the grounds of the US Army Quartermaster Intermediate Depot near Norfolk on February 21, 1922. The box-kite rudder has jammed, causing the airship to begin a rapid descent, upon which it struck power lines, causing the flammable hydrogen within the airship's envelope to explode. (Courtesy of Hampton History Museum, via NASA Langley Research Center.)

Roma Crash Site. This aerial view shows the *Roma* crash site on the grounds of the US Army Quartermaster Intermediate Depot near Norfolk. The airship struck the power lines visible at lower center. The frames of the airship's box kite elevator and rudder remain snagged by the telephone pole near the bend in the road. A total of 34 crewmen lost their lives in what was America's worst air disaster at the time. (Courtesy of Martin Copp, via NASA Langley Research Center.)

AIRSHIP/AIRCRAFT PARASITE STUDIES. In September 1923, precedent-setting flight experiments were successfully performed by airships and Sperry Messenger biplane aircraft in the skies over Langley Field. The studies effectively demonstrated the feasibility of the "aerial aircraft carrier" concept. (Courtesy of USAF, Air Combat Command History Office.)

MAY 1923 VIEW OF THE LANGLEY FIELD AERO REPAIR SHOP. Four biplanes are visible within the facility: from left to right are a de Havilland DH-4, Sperry Messenger, Boeing MB-3, and another DH-4 behind the Boeing. (Courtesy of USAF, Air Combat Command History Office.)

August 1923 Inspection of the Field by Gen. Mason Patrick. Visible in the background are Martin MB-2 and de Havilland DH-4 bombers. (Courtesy of US National Archives, Still Pictures Branch.)

January 1924 Around-the-World Flight Training. In January 1924, aircrews trained at Langley Field in the prototype Douglas World Cruiser *Boston II* for the famous around-the-world flight. Visible below the aircraft is the Old North King Street bridge. (Courtesy of USAF, Air Combat Command History Office.)

BOMBERS EMBARK ON FLIGHT TO THE NATION'S CAPITAL. In 1924, 2nd Bomb Group Martin MB-2 bombers flew from Langley Field to Washington, DC. Visible on the shore of the Back River are the seaplane hangars, presently the site of the Langley Yacht Club. Also visible are the original headquarters building for NACA (at right, behind tennis courts) and the two original NACA flight research hangars (dark-colored hangars at lower left). (Courtesy of US National Archives, Still Pictures Branch.)

LANGLEY FIELD FLIGHT LINE DURING 1925 INSPECTION. This aerial view shows the flight line during an inspection by top military brass in April 1925. Note the original two NACA flight research hangars (dark-colored roofs) at far right, the lineup of NACA flight research subjects, and a huck starter (aircraft engine starting vehicle) parked at the NACA hangars' entrance. (Courtesy of US National Archives, Still Pictures Branch.)

1st Provisional Air Brigade Headquarters. This view of the 1st Provisional Air Brigade Headquarters at Langley Field was taken during war maneuvers on October 15, 1925. (Courtesy of US National Archives, Still Pictures Branch.)

1st Provisional Air Brigade Staff. The 1st Provisional Air Brigade staff poses for a publicity photograph during war maneuvers at Langley Field on October 15, 1925. From left to right are Brig. Gen. James E. Fechet, Maj. Carl Spatz, Major Jouett, Capt. R.H. Wooten, Capt. Ira Eaker, and Lt. B.R. Dallas. (Courtesy of US National Archives, Still Pictures Branch.)

1st Provisional Air Brigade Pursuit Aircraft. Pursuit biplanes are pictured on the flying field during war maneuvers at Langley Field on October 15, 1925. Note the Fokker D.VII at center left. The Dutch aircraft has been retrofitted with an American engine. (Courtesy of US National Archives, Still Pictures Branch.)

1st Provisional Air Brigade Attack Aircraft in Flight. DH-4 attack aircraft of the 1st Provisional Air Brigade are pictured in flight near the Army Air Service Tactical School Operations Office during war maneuvers on October 15, 1925. (Courtesy of US National Archives, Still Pictures Branch.)

1st Provisional Air Brigade Aircraft. Pursuit, attack, and bomber aircraft participate in war maneuvers on October 15, 1925, at Langley. Visible are British-built SE-5 pursuit aircraft and DH-4 attack aircraft, a Fokker D.VII with retrofitted American engine pursuit aircraft, Thomas-Morse MB-3 pursuit aircraft, Boeing PW-9 pursuit aircraft, and Martin MB-2 bomber. (Courtesy of US National Archives, Still Pictures Branch.)

Martin Bombers in Flight. A formation of Langley Field–based Martin bombers is pictured on October 16, 1925. (Courtesy of US National Archives, Still Pictures Branch.)

GUNNERY RANGE. This aerial view shows the Langley Field machine gun range at Plum Tree Island with targets visible on October 28, 1925. (Courtesy of US National Archives, Still Pictures Branch.)

CURTISS JENNY TARGET GLIDER MOTHERSHIP. A Curtiss Jenny is pictured at Langley Field in late 1925. The Jenny mothership air-released a glider that served as a target for pursuit aircraft pilots and rear gunners of attack aircraft. (Courtesy of USAF, Air Combat Command History Office.)

DH-4 Performing Gunnery Practice. The rear gunner of a de Havilland DH-4 performs gunnery practice on a towed target on October 27, 1925. (Courtesy of US National Archives, Still Pictures Branch.)

A Wintry Scene. Langley Field is shown here after a March 2, 1927, snowstorm. (Courtesy of US National Archives, Still Pictures Branch.)

Massive Snow Accumulation. The heavy snowfall from the winter storm of March 2, 1927, is evident in this Langley Field scene. (Courtesy of US National Archives, Still Pictures Branch.)

Keystone and Martin Bombers. Keystone and Martin bombers of the 2nd Bomb Group fly in formation during maneuvers on May 17, 1928. (Courtesy of US National Archives, Still Pictures Branch.)

Keystone and Martin Bombers as Seen from the Ground. Keystone and Martin bombers of the 2nd Bomb Group are seen flying in formation during maneuvers on April 28, 1931. (Courtesy of US National Archives, Still Pictures Branch.)

Aerial View of the Newly Constructed Post Chapel. The new post chapel at Langley Field is seen from the air on April 23, 1935. (Courtesy of US National Archives, Still Pictures Branch.)

Martin B-10 Bombers Performing Winter Maneuvers. Martin B-10 bombers take off from a snowy Langley Field on January 30, 1936, to perform winter maneuvers. (Courtesy of US National Archives, Still Pictures Branch.)

Martin B-10Bs on the Flight Line. Bombers are pictured on Langley Field's flight line on June 15, 1936. (Courtesy of US National Archives, Still Pictures Branch.)

Prototype B-17 Flying Fortress. On March 1, 1937, the prototype of the famous Boeing B-17 Flying Fortress bomber series, the YB-17, arrived at Langley Field. (Courtesy of US National Archives, Still Pictures Branch.)

Aerial View of the YB-17 Arrival. A ceremony is held at Langley Field celebrating the arrival of the YB-17 on March 1, 1937. (Courtesy of US National Archives, Still Pictures Branch.)

YB-17 AIRCREW AND US ARMY AIR CORPS BRASS. The aircrew of the YB-17 and US Army Air Corps brass prepare to meet the press during the aircraft's arrival ceremonies. (Courtesy of US National Archives, Still Pictures Branch.)

RECEIVING A HEARTY WELCOME. The YB-17 pilot, Maj. Barney M. Giles, is greeted by Lt. Gen. Frank M. Andrews, general headquarters commander. (Courtesy of US National Archives, Still Pictures Branch.)

THE AMAZING BOEING XB-15 LONG-RANGE HEAVY BOMBER. On February 14–15, 1939, the Boeing XB-15 bomber achieved an aviation milestone when it successfully transported 3,250 pounds of medical supplies from Langley Field to Chile to provide relief following a massive earthquake. The historic airlift mission was performed nonstop and lasted over 29 hours. (Courtesy of USAF.)

8TH PURSUIT GROUP LINEUP. On May 20, 1939, several 8th Pursuit Group Curtiss P-36A Hawk fighters, a Martin B-10B bomber, and a Douglas O-43 observation/scout aircraft are pictured on the flight line at Langley Field. (Courtesy of US National Archives, Still Pictures Branch.)

36TH PURSUIT GROUP CURTISS P-36A HAWK LINEUP. P-36A fighters of the 36th Pursuit Group line up at Langley Field in 1940. (Courtesy of USAF.)

33RD PURSUIT SQUADRON, 8TH PURSUIT GROUP CURTISS P-40 FIGHTERS. P-40s of the 33rd Pursuit Squadron, 8th Pursuit Group are pictured at Langley Field in 1941. (Courtesy of USAF.)

Two

Winning a World War 1942–1945

The 1st Sea Search Attack Group was formed at Langley Field on June 8, 1942, to pioneer revolutionary new aerial anti-submarine warfare (ASW) technologies. Flying modified Douglas B-18B Bolo bombers, the aerial ASW air unit pioneered the magnetic anomaly detector (MAD) boom, which is now a standard feature on all ASW patrol aircraft the world over. The MAD systems pioneered by the 1st Sea Search Attack Group included the Mark IV-B2, Mark VI, and Mark X MAD systems. They were effectively outfitted on both US Army Air Forces (USAAF) and US Navy patrol aircraft, resulting in the sinking of numerous German U-boats in the Atlantic and in Europe. Langley served primarily as a bomber base during World War II, and pioneering radar bombing studies were also conducted there during the latter portion of the war.

The 375th and 376th Fighter Squadrons, which flew the new Republic P-47 Thunderbolt fighter-bomber, were also based at Langley from May through July 1943. During their brief assignment to Langley Field, these fighter squadrons helped to train new Thunderbolt pilots and maintenance crews. The P-47 went on to become a USAAF fighter-bomber mainstay in both the European and Pacific theaters during World War II.

ATTACK BOMBER MAINTENANCE. Ground crews perform maintenance on a Douglas A-20 Havoc attack bomber at Langley Field in July 1942. (Courtesy of US Library of Congress, Alfred T. Palmer.)

Side View of Ground Crews Performing Maintenance. A Douglas A-20 is receiving maintenance from ground crews at Langley Field in July 1942. Note that some of the crew are preparing to load two practice bombs in the bomb bay of the aircraft. (Courtesy of US Library of Congress, Alfred T. Palmer.)

Sun Setting on a Flying Fortress. An early-war B-17 Flying Fortress and its crew are pictured at sunset at Langley Field in July 1942. (Courtesy of US Library of Congress, Alfred T. Palmer.)

Flying Fortress Aircrew. An early-war B-17 Flying Fortress aircrew discusses its flight mission prior to taking off from Langley Field in 1942. (Courtesy of US Library of Congress, Alfred T. Palmer.)

YB-17 Flying Fortress Mission Prep. A YB-17 Flying Fortress is prepped for a mission at Langley Field in 1942. (Courtesy of US Library of Congress, Alfred T. Palmer.)

YB-17 Revving Up Engines. A YB-17 Flying Fortress revs up its engines at Langley Field in 1942. (Courtesy of US Library of Congress, Alfred T. Palmer.)

MAD-Equipped Douglas B-18 on an ASW Mission. A Langley Field–based Douglas B-18B ASW patrol bomber trails the submarine detector, concealed in a "towed bird," while performing an ASW mission over the Atlantic Ocean. (Courtesy of USAF, Air Combat Command History Office.)

LANGLEY FIELD–BASED CONSOLIDATED B-24D ASW PATROL BOMBER. A Consolidated B-24D ASW patrol bomber is refueled in 1943. The aircraft was equipped with an anti-shipping radar dome, visible behind the bomb bay doors. (Courtesy of Martin Copp, via NASA Langley Research Center.)

Experimental Radar Bombing Boeing B-17F Flying Fortresses. These experimental B-17 Flying Fortresses were used at Langley Field in experimental radar bombing studies and training from 1944 to 1945. (Courtesy of Martin Copp, via NASA Langley Research Center.)

Three

Communism and Quelling Cold War Tensions 1946–1962

On May 25, 1946, the Tactical Air Command (TAC) Headquarters was created at Langley Field. TAC was established to organize, train, outfit, and maintain combat air units that could be deployed on short notice in times of peace and war. As of August 1947, USAF TAC squadrons based at Langley included the 160th Tactical Reconnaissance Squadron, which flew RF-80 Shooting Stars; the 161st Tactical Reconnaissance Squadron, which also operated RF-80s; and the 12th Tactical Reconnaissance Squadron, which flew RA-26s. In January 1948, Langley Field was renamed Langley Air Force Base, following the creation of the US Air Force in 1947. As of April 1949, the Continental Air Command at Langley comprised the 4th Fighter Wing, which flew Lockheed F-80 Shooting Stars and North American F-86A Sabres.

LAFB played a critical role supporting US and UN forces against the communist forces of North Korea, China, and the Soviet Union by providing USAF fighter pilots with tactical training in the service's premier jet fighter, the North American F-86 Sabre. As of March 1951, the 47th Bomber Wing's Douglas A-26 Invader attack/bomber aircraft and North American B-45 Tornado jet bomber aircraft (America's first jet bomber) were stationed at LAFB. In May 1953, A-26s, Martin B/RB-57 Canberra bombers, and new supersonic North American F-100 Super Sabre jet fighter/bombers assigned to the 405th Fighter-Bomber Wing were stationed at LAFB. In 1958, Boeing KB-29 and KB-50 long-range aerial refueling aircraft assigned to the 450th Air Refueling Wing were stationed at LAFB.

During the Cuban Missile Crisis in October 1962, USAF Strategic Air Command (SAC) Boeing B-47A Stratojet bombers, which had nuclear weapons–carrying capability, were detached to LAFB to conduct carpet bombings of Cuba should the crisis escalate. Fortunately, the Soviet Union backed down and removed its nuclear missiles from Cuba.

New Flight Operations Control Tower. This 1946 photograph of Langley Field shows a new flight operations control tower in service. Also visible in the background is the dirigible hangar, which was torn down in 1947. A massive parking lot now covers the site. (Courtesy of US National Archives, Still Pictures Branch.)

Some of the First Production Lockheed P-80 Shooting Star Fighters. The P-80 was a USAF mainstay during the Korean War. These were based at LAFB during the late 1940s. (Courtesy of USAF.)

Korean War Warrior. This North American F-86A Sabre jet fighter was part of the 4th Fighter Wing, which served at LAFB from May 1949 through August 1950. The Sabre jet went on to become a tremendous air superiority fighter for the USAF during the Korean War, totally outclassing its competition, the Soviet-built Mikoyan-Gurevich MiG-15 flown by North Korea and China. (Courtesy of Martin Copp, via NASA Langley Research Center.)

North American B-45 Tornado Lineup. These B-45A-5-NA Tornadoes of the 47th Light Bomb Wing are pictured on the flight line at LAFB in 1952. The Tornado was America's first jet bomber. (Courtesy of USAF.)

LAFB-Based Republic F-84F-35-RE Thunderstreak. A Thunderstreak of the 405th Fighter-Bomber Wing is pictured in flight in 1955. (Courtesy of USAF.)

1st Tactical Fighter Wing (TFW) Predecessor. The 94th Fighter-Interceptor Squadron was a predecessor fighter squadron to the 1st TFW. One of the 94th's North American F-86D Sabre Dogs is pictured at Selfridge Air Force Base, Michigan, in 1955. The 1st TFW later served as a fighter wing host and mainstay at LAFB. (Courtesy of USAF.)

The "Huns." During the mid-1950s, the 405th Fighter-Bomber Wing, based at LAFB, flew F-100C Super Sabres, later affectionately dubbed "Huns" during the Vietnam War. Super Sabres performed countless close air support sorties for US Army and Marine ground forces during the conflict. (Courtesy of USAF.)

Convair F-102A Delta Dagger. In 1960, the 48th Fighter Interceptor Squadron (FIS), based at LAFB, operated the Convair F-102A Delta Dagger supersonic interceptor. (Courtesy of USAF.)

Boeing B-47E Stratojet. A Boeing B-47E Stratojet is pictured at LAFB during the Cuban Missile Crisis of October 1962. During the crisis, numerous Stratojets were deployed to LAFB and would have carpet bombed Cuba had tensions escalated. (Courtesy of USAF.)

Four

US Efforts in Vietnam and the Post-Vietnam Era

1963–1980

From 1963 to 1975, LAFB hosted the 463rd Troop Carrier Wing (1963), 316th Tactical Airlift Wing (1966), and the Military Airlift Command's (MAC) 316th Tactical Airlift Wing (1975). All the units stationed at LAFB flew C-130 Hercules aircraft in support of America's involvement in the Vietnam War.

In October 1965, C-130s from the 463rd TCW took part in a massive airdrop into Vietnam. The LAFB-based unit helped drop 1,200 South Vietnamese paratroopers just north of Saigon in an attempt to root out Viet Cong troops. In April 1966, a new, $1.5-million TAC photo reconnaissance laboratory became operational at LAFB.

To meet global air superiority needs, the 1st Tactical Fighter Wing was moved from MacDill Air Force Base, Florida, to LAFB in January 1976. The unit was the first USAF unit to receive the new McDonnell Douglas F-15A Eagle. While at MacDill Air Force Base, 1st TFW predecessors, the 4424th Combat Crew Training Squadron (CCTS) and the 94th Tactical Fighter Squadron (TFS), trained Martin B-57G Canberra night intruder jet bomber and McDonnell Douglas F-4E Phantom II air superiority fighter-bomber aircrews, who later flew combat sorties during the Vietnam War.

LOCKHEED C-130E HERCULES. Visible in this mid-1960s photograph of the flight line at LAFB are 316th Troop Carrier Wing C-130Es (center right). LAFB-based C-130s played a critical role in the mobilization of US and South Vietnamese troops, vehicles, weaponry, and equipment during the Vietnam War. (Courtesy of USAF, Air Combat Command History Office.)

LAFB-BASED CONVAIR F-106 DELTA DARTS. LAFB-based 48th FIS Convair F-106 Delta Dart supersonic jet interceptors are pictured in flight in 1969. The F-106 replaced the F-102 in the supersonic jet interceptor role for the USAF. (Courtesy of USAF.)

New TAC Headquarters. On April 27, 1971, a new TAC office headquarters building was opened at LAFB. Visible in this aerial photograph are the TAC headquarters (foreground) and LAFB Yacht Club (background). (Courtesy of US National Archives, Still Pictures Branch.)

New Base Hospital. On April 21, 1971, a new base hospital opened at LAFB. Visible here are the hospital's main entrance and parking lot as well as the base exchange (BX) service station (right). (Courtesy of US National Archives, Still Pictures Branch.)

MARTIN B-57G NIGHT INTRUDER BOMBER. This Martin B-57G was assigned to MacDill Air Force Base's 4424th Combat Crew Training Squadron (CCTS) during the early 1970s. The 4424th CCTS was a predecessor to the 1st TFW. (Courtesy of USAF.)

McDONNELL DOUGLAS F-4E PHANTOM II AIR SUPERIORITY FIGHTER-BOMBER. This F-4E Phantom II was assigned to the 94th Tactical Fighter Squadron (TFS) out of MacDill Air Force Base, Florida, in the early 1970s. The 94th TFS was a predecessor of the 1st TFW. (Courtesy of USAF.)

1st TFW Receives Its First McDonnell Douglas F-15A Eagle. The 1st TFW, the first operational F-15 squadron, received its first F-15A air superiority fighter at LAFB in January 1976. (Courtesy of USAF.)

Fairchild A-10 Thunderbolt II Ground-Attack Aircraft Debut. The initial TAC combat-operational Fairchild A-10 Thunderbolt II ground-attack aircraft makes its debut during an air show at LAFB on March 20, 1976. (Courtesy of US National Archives, Still Pictures Branch.)

EC-135 Command and Control and F-15 Refueling Aircraft. The 6th Airborne Command and Control Squadron flew these EC-135 command and control aircraft and F-15 refueling aircraft. They are pictured on the tarmac at LAFB during the late 1970s. (Courtesy of USAF, Air Combat Command History Office.)

Bell UH-1 Support Helicopter. A 1st TFW UH-1 support helicopter lifts off from LAFB during the late 1970s. (Courtesy of USAF, Air Combat Command History Office.)

Five

"Peace through Strength" and the Langley Buildup 1981–1989

The 1980s and the "Peace through Strength" policy approach to national defense embraced by Pres. Ronald Reagan and his administration led to a great military buildup both nationally and at LAFB. Major base renovations and improvements commenced in February 1980 with the opening of a $4-million commissary. An enhanced gatehouse and visitor center opened in 1982. The Langley Memorial Park, featuring landmark F-86 (Korean War), F-105 (Vietnam War), and F-15 (Cold War) displays, was unveiled in 1983. In December 1985, a state-of-the-art flight operations control tower became operational at LAFB. In 1987, a unique, remote LAFB primary care clinic came online at nearby Bethel Manor.

In 1981, the 48th FIS transitioned from the venerable F-106 to the advanced F-15, and in January 1982, a 1st Tactical Fighter Wing F-15 Eagle became the first aircraft of its type to intercept a Soviet Tu-95 "Bear" long-range bomber. In 1986, the CONUS (Continental United States) North American Aerospace Defense Command (NORAD) Region (CONR) and the CONUS NORAD Region Operations Control Center (ROCC) (Headquarters, First Air Force) became operational at LAFB.

Olympic Torch Arrives. On January 31, 1980, Maj. Gen. Carl H. Cathey, deputy chief of staff, delivers a welcoming speech to visitors at an Olympic torch ceremony held at LAFB. Sandy Norris, a XIII Winter Olympic Games torchbearer, can be seen in the yellow suit. The flame was transferred, via run, from LAFB to Lake Placid, New York. (Courtesy of US National Archives, Still Pictures Branch.)

Yorktown Bicentennial Eagles. A pair of LAFB-based 1st TFW F-15 Eagles embark on a Yorktown bicentennial celebratory flyover on October 13, 1981. (Courtesy of US National Archives, Still Pictures Branch.)

LAFB-Based 48th FIS F-15 Eagles. A pair of F-15 Eagles, with a full complement of AIM-9 Sidewinder (on wing pylons) and AIM-7 Sparrow air-to-air missiles, is pictured in flight on July 1, 1982. (Courtesy of US National Archives, Still Pictures Branch.)

Air-to-Air Missile Test. A 48th FIS F-15 launches an AIM-7 Sparrow air-to-air missile during a weapons test on September 1, 1982. (Courtesy of US National Archives, Still Pictures Branch.)

"Loaded Up for Bear." A 1st TFW F-15 is pictured in flight on June 15, 1983, loaded up with AIM-9 Sidewinder and AIM-7 Sparrow air-to-air missiles. (Courtesy of US National Archives, Still Pictures Branch.)

New Flight Operations Control Tower. Construction progresses on a new flight operations control tower at LAFB on June 7, 1984. (Courtesy of US National Archives, Still Pictures Branch.)

TAC Commander in Chief's Home. The home of the TAC commander in chief is pictured on June 14, 1984. (Courtesy of US National Archives, Still Pictures Branch.)

Air Defense TAC Building. The Air Defense TAC Building at LAFB is pictured on July 21, 1984. (Courtesy of US National Archives, Still Pictures Branch.)

Bio/Chemical Threat Exercise. LAFB guards, donning special protective gear, take up positions in front of the 1st TFW Headquarters for an operational readiness inspection on March 28, 1985. (Courtesy of US National Archives, Still Pictures Branch.)

Exercise Solid Shield 85 Preparations. A trio of 1st TFW F-15A Eagles is serviced in preparation for Exercise Solid Shield 85 operations on April 30, 1985. (Courtesy of US National Archives, Still Pictures Branch.)

Exercise Solid Shield 85 Control Tower Activity. This view inside the LAFB control tower was taken during Exercise Solid Shield 85 on April 30, 1985. (Courtesy of US National Archives, Still Pictures Branch.)

***Spirit of Yorktown* in Flight.** A 1st TFW F-15 Eagle, dubbed *Spirit of Yorktown*, flies over the Yorktown Monument and the waterfront of Yorktown, Virginia, on August 5, 1986. (Courtesy of US National Archives, Still Pictures Branch.)

Eagles over the Newport News Shipyard. Four 1st TFW F-15 Eagles fly over Newport News Shipbuilding in Newport News, Virginia, on August 5, 1986. (Courtesy of US National Archives, Still Pictures Branch.)

***Spirit of Hampton* in Flight.** The 1st TFW F-15 Eagle *Spirit of Hampton* flies over the Hampton Coliseum on August 5, 1986. (Courtesy of US National Archives, Still Pictures Branch.)

Loading a Sparrow Missile on an F-15. Ground crews load an AIM-7 Sparrow missile on an F-15 Eagle with the assistance of an M-1 loader at LAFB on October 11, 1988. (Courtesy of US National Archives, Still Pictures Branch.)

Arming an F-15. S.Sgt. Steven Delacruz (right) and Amn. Charley Visele of the 1st Aircraft Generation Squadron, based at LAFB, arm an F-15 Eagle with an AIM-7 Sparrow missile on October 11, 1988. (Courtesy of US National Archives, Still Pictures Branch.)

F-15 Eagle in Flight. A 1st TFW F-15 Eagle, armed with four Sidewinder short-range, heat-seeking, air-to-air missiles, banks to the right on a flight during the late 1980s. (Courtesy of US National Archives, Still Pictures Branch.)

Six

Langley in an Era of Regional Conflicts 1990–2000

Iraq invaded Kuwait on August 2, 1990, and occupied that nation for seven months. Shortly after this act of aggression, the United States and its partner nations began amassing a formidable military force to overwhelm Saddam Hussein's army and eject the Iraqi forces from Kuwait. The military buildup assembled in the deserts surrounding Iraq and Kuwait became known as Operation Desert Shield. A potential military encounter with Iraq provided US defense planners with a golden opportunity to test and evaluate one of America's newest and most advanced weapons, the Lockheed F-117A Nighthawk (a stealth fighter that actually performed as a bomber). In August, the Pentagon deployed a squadron of the new stealth fighters from the Tonopah Test Range, Nevada, to an air base in Saudi Arabia. On their way to the Middle East, the Nighthawks stopped at LAFB for refueling and service. These same Nighthawks spearheaded the initial Allied Coalition bombing missions against Iraq during Operation Desert Storm, which commenced on January 16, 1991.

On August 7–8, 1990, the LAFB-based 71st and 27th Tactical Fighter Squadrons were deployed to Dhahran, Saudi Arabia, to serve as important initial elements of Operation Desert Shield. Both squadrons conducted initial combat sorties on January 17, 1991. During Operation Desert Storm, a 71st TFS F-15 downed a French-built Iraqi Mirage F-1 in aerial combat. Throughout the 1990s and into the early 2000s, LAFB-based F-15s performed countless patrol sorties of the no-fly zones over Iraq during Operations Southern and Northern Watch.

On August 12, 1990, the US Central Command Air Forces (USCENTAF) TAC battle staff was formed from elements of the TAC headquarters battle staff and USCENTAF REAR battle staff and established at LAFB. Langley's host air wing was renamed the 1st Fighter Wing on October 1, 1991. On June 1, 1992, TAC headquarters was inactivated and Air Combat Command (ACC) established at LAFB. The effects of Russia's glasnost policy were experienced at LAFB from July 14 to 18, 1992, when two Sukhoi Su-27 "Flanker" fighters and one Il-76 "Candid" transport assigned to the 91st Research Instructor Fighter Aviation Regiment (Lipetsk Air Base), Frontal Aviation Division, Russian Air Force, made a goodwill visit to LAFB. In December 1997, a $9-million, state-of-the-art civil engineering building was erected at LAFB.

On an Important Mission. In August 1990, F-117A Nighthawk stealth fighters of the 37th TFW, from Tonopah Test Range, Nevada, made an overnight stop at LAFB before embarking on a deployment to Saudi Arabia in support of Operation Desert Shield. (Both, courtesy of USAF.)

LAFB-Based 1st TFW F-15C Eagle during Operation Desert Shield. This F-15C Eagle is pictured at an airbase in Saudi Arabia during Operation Desert Shield. Note the Patriot surface-to-air missile batteries visible behind the aircraft revetments. (Courtesy of USAF.)

1st TFW F-15C Eagle Operations. A 1st TFW F-15 remains parked in a makeshift hangar while a second F-15 heads out on a patrol flight during Operation Desert Shield. (Courtesy of US National Archives, Still Pictures Branch.)

Guarding 1st TFW F-15C Eagles during Operation Desert Shield. Two USAF security policemen guard 1st TFW F-15s. (Courtesy of US National Archives, Still Pictures Branch.)

ACC Establishment Ceremony. Former TAC commander Gen. John Loh gives a speech at the ACC establishment ceremony on June 1, 1992. (Courtesy of US National Archives, Still Pictures Branch.)

Lockheed C-5 Galaxy. On January 27, 1995, this Lockheed C-5 Galaxy heavy transport, based at Travis Air Force Base, California, took part in a four-day joint military service proficiency exercise at LAFB. (Courtesy of US National Archives, Still Pictures Branch.)

General Dynamics EF-111 Raven. On January 31, 1995, this and several other General Dynamics EF-111 electronic warfare aircraft assigned to the 429th Electronic Combat Squadron at Cannon Air Force Base, New Mexico, paid a visit to LAFB. EF-111s effectively participated in Operation Desert Storm. (Courtesy of US National Archives, Still Pictures Branch.)

General Dynamics F-111 Aardvark. On January 31, 1995, this and several other swing-wing General Dynamics F-111 Aardvark fighter-bombers assigned to the 523rd Fighter Squadron at Cannon Air Force Base also visited LAFB. F-111s were used during the Vietnam War, Operation El Dorado Canyon (a retaliatory attack on Libya), and Operation Desert Storm. (Courtesy of US National Archives, Still Pictures Branch.)

Boeing E-3 Sentry Early Warning/Command Control Aircraft. Based at Tinker Air Force Base, Oklahoma, this Boeing E-3B Sentry Airborne Early Warning Control System (AWACS) heads out on a support mission for the 27th Wing on January 31, 1995. Note the NASA Langley Research Center Flight Research Hangar/Building in the background. (Courtesy of US National Archives, Still Pictures Branch.)

Loading Mobile Equipment aboard a C-17. USAF staff sergeant Day helps load mobile equipment aboard a C-17 Globemaster bound for Saudi Arabia at LAFB on August 24, 1995. (Courtesy of US National Archives, Still Pictures Branch.)

McDonnell Douglas C-17 Globemaster Transport. A C-17 remains parked at LAFB and ready to load equipment, supplies, and personnel belonging to the 74th Air Control Squadron for airlift to Saudi Arabia on August 24, 1995. (Courtesy of US National Archives, Still Pictures Branch.)

Advanced Echelon (ADVON) 1st Fighter Wing Deployment to Jordan. On March 30, 1996, the 1st Fighter Wing ADVON team arrived at an air base in Jordan via a C-141B Starlifter transport to assist the USAF and multinational forces in carrying out routine patrols of the no-fly zone above southern Iraq. (Courtesy of US National Archives, Still Pictures Branch.)

1st Combat Communications Squadron in Jordan. On April 11, 1996, the LAFB-based 1st Combat Communications Squadron established a base of operations at Azaraq Air Base, Jordan, to assist routine patrols of the no-fly zone above southern Iraq. (Courtesy of US National Archives, Still Pictures Branch.)

Patrolling the No-Fly Zone above Iraq. A LAFB-based 1st TFW F-15 Eagle patrols the no-fly zone above Iraq during the late 1990s. (Courtesy of US National Archives, Still Pictures Branch.)

National Security Readiness Subcommittee Hearing. On March 3, 1997, the House of Representatives National Security Readiness Subcommittee held a hearing at Quesada Hall at LAFB to discuss and assess the preparedness and quality of military life. The committee was led by Congressman Herb Bateman of Virginia, and the hearing was attended by senior military leaders. (Courtesy of US National Archives, Still Pictures Branch.)

The Unique Air-Transportable Hospital. The Air-Transportable Hospital was developed by the LAFB ACC in September 1997 for erection at training exercises and inspections. (Courtesy of US National Archives, Still Pictures Branch.)

LAFB Weapons Team Members at Enhanced Laser-Guided Bomb Feasibility Test. On March 9, 1999, LAFB weapons team members met with representatives of Raytheon Missile Systems and Lockheed Martin at Luke Air Force Base, Arizona, to conduct a weapon/aircraft feasibility test between the new Enhanced Guided Bomb Unit-27 and an F-16 Fighting Falcon fighter. (Courtesy of US National Archives, Still Pictures Branch.)

ACC Commander Meets the Commander of the German Air Force Command. On July 29, 2000, USAF general John Jumper (center), ACC commander, LAFB, meets with German lieutenant general Peter Vogler (left), commander, German Air Force Command, and Lt. Col. Rob Topp, commander, 63rd Fighter Squadron, Luke Air Force Base, Arizona. The trio stand next to a Luftwaffe F-104 Starfighter wing exhibited at the 63rd Fighter Squadron headquarters. (Courtesy of US National Archives, Still Pictures Branch.)

LAFB-Based USAF 1st Security Forces Squadron Contending Warrior Team. The USAF 1st Security Forces Squadron Contending Warrior Team practices securing the landing zone in training maneuvers conducted at LAFB on September 9, 2000, in preparation for the Contending Warrior Competition that would be held later in the desert badlands of Nevada. (Courtesy of US National Archives, Still Pictures Branch.)

Seven

The War on Terror, Iraq, and the Future

2001–Present

On September 11, 2001, two LAFB-based Air National Guard F-16 Fighting Falcons took off from Langley to patrol the skies above Washington, DC, and shoot down any further hijacked airliners that posed a threat to the American capital following the attacks on the World Trade Center in New York City and the Pentagon. Unfortunately, the Fighting Falcons were unable to reach the air space above the Pentagon in time to prevent the tragedy that occurred there. Throughout the remainder of the first decade of the 2000s, LAFB-based F-15s participated in Operation Noble Eagle, which involved routine anti-air-based terrorist threat patrol sorties over Washington and the East Coast.

LAFB-based F-15s continued patrolling the no-fly zone in Iraq during the early 2000s and into Operation Iraqi Freedom in 2003. On May 12, 2005, the LAFB-based 1st Fighter Wing became the first USAF squadron to operate the new, stealthy Lockheed Martin F-22A Raptor air superiority fighter.

On January 7, 2010, elements from the Army base at Fort Eustis and LAFB were combined to form Joint Base Langley-Eustis (JBLE). The newly formed 633rd Air Base Wing was designated to command joint base operations at LAFB.

On September 22, 2014, the F-22A Raptor made its combat debut as four JBLE-based 27th Fighter Squadron F-22s attacked terrorist forces in Syria.

From December 2 to 18, 2015, JBLE (1st Fighter Wing) served as the home base for a trilateral exercise, consisting of USAF, British Royal Air Force (RAF), and French Air Force aircraft and personnel elements.

On July 22, 2016, JBLE celebrated the 95th anniversary of the bombing and sinking of the *Ostfriesland* by hosting and orchestrating a mock battleship bombing in which seven air wings, operating eight different types of aircraft, performed the role of the 1st Provisional Air Brigade in 1921 and dropped dummy bombs over the *Ostfriesland*'s ocean gravesite. The event helped to serve notice that JBLE, one of the oldest Air Force bases in the United States, remains primed to carry the nation into the future on the leading edge of Air Force airpower.

Welcoming Pres. George W. Bush and First Lady Laura Bush. Colonel Goldfein, commander of the 1st Fighter Wing, greets Pres. George W. Bush and First Lady Laura Bush at LAFB on March 4, 2001. The Bushes visited LAFB on their way to attending the christening ceremony of the USS *Ronald Reagan* (CVN-76) at the Newport News shipyard. (Courtesy of US National Archives, Still Pictures Branch.)

Practicing Putting Out an Aircraft Fire. USAF sergeant Ronald Ackerson (right) and Amn. Jason Reese, 1st Civil Engineering Squadron firefighters, practice putting out an aircraft fire within the "Fire Pit" at LAFB during the 1st Fighter Wing Operational Readiness Inspection (ORI) on March 19, 2001. (Courtesy of US National Archives, Still Pictures Branch.)

USAF Global Hawk Unmanned Aerial Vehicle (UAV). On June 21, 2001, the USAF Global Hawk UAV (drone) arrived at LAFB after completing a nonstop transcontinental flight from Edwards Air Force Base, California. The UAV participated in a NATO seminar conducted by the Supreme Allied Commander, Atlantic (SACLANT). (Courtesy of US National Archives, Still Pictures Branch.)

Lancer Display. A Rockwell/Boeing B-1B Lancer, a strategic nuclear swing-wing bomber, was on display at LAFB as part of the SACLANT seminar on June 21, 2001. (Courtesy of US National Archives, Still Pictures Branch.)

Stratofortress Display. A Boeing B-52 Stratofortress strategic nuclear bomber was also on display as part of the SACLANT seminar. (Courtesy of US National Archives, Still Pictures Branch.)

Operation Noble Eagle Lockheed Martin F-16 Fighting Falcons. On February 1, 2002, as part of Operation Noble Eagle, F-16s of the 148th Fighter Wing, Minnesota Air National Guard, were assigned to LAFB to conduct combat air patrols (CAPs) above heavily populated cities on the East Coast. (Courtesy of US National Archives, Still Pictures Branch.)

Operation Northern Watch No-Fly Zone Patrol. On October 10, 2002, a 20th Fighter Wing F-16C Fighting Falcon, based at Shaw Air Force Base, South Carolina, and armed with high-speed anti-radiation missiles (HARMs), flies with two LAFB-based 1st Fighter Wing F-15C Eagles on an Operation Northern Watch no-fly zone patrol above northern Iraq. (Courtesy of US National Archives, Still Pictures Branch.)

Operation Iraqi Freedom Lockheed Martin F-117A Nighthawk. On April 14, 2003, USAF colonel Steve Miller, commander of the 1st Fighter Wing, and other staff members welcome the pilot of a 49th Fighter Wing F-117A Nighthawk based at Holloman Air Force Base, New Mexico, upon their return stop at LAFB from Operation Iraqi Freedom. Note the successful mission completion tally markings on the aircraft, denoted by individual nighthawk icons. (Courtesy of US National Archives, Still Pictures Branch.)

Operation Iraqi Freedom 1st Fighter Wing F-15s Prepare for Flight Home. LAFB-based F-15C Eagles prepare to head home following the successful completion of their mission in Operation Iraqi Freedom on April 22, 2003. (Courtesy of US National Archives, Still Pictures Branch.)

A Patriotic Homecoming. USAF families and supporters, gathered together at LAFB, provide a patriotic welcome home to 71st Fighter Squadron, 1st Fighter Wing F-15s after the completion of their 71-day Operation Iraqi Freedom mission on April 24, 2003. (Courtesy of US National Archives, Still Pictures Branch.)

A Heartwarming Welcome Home for Dad. Capt. Michael Moran, an F-15 pilot with the LAFB-based 71st Fighter Squadron, is the recipient of a heartwarming welcome from sons Colton and Blake on April 24, 2003. (Courtesy of US National Archives, Still Pictures Branch.)

Hurricane Isabel Pummels the LAFB Marina. The full fury of Hurricane Isabel is felt at the LAFB marina on September 18, 2003. (Courtesy of US National Archives, Still Pictures Branch.)

Hurricane Isabel Wind and Debris Damage. Hurricane Isabel wreaked extensive flood and wind damage that totaled millions of dollars. Here, cleanup crews remove fallen trees and debris from the premises on LAFB on September 22, 2003. (Courtesy of US National Archives, Still Pictures Branch.)

Predator UAV. On October 20, 2003, ACC civilian leaders were given the opportunity for a glimpse of two effective weapons in fighting the War on Terror, the Global Hawk and Predator UAVs. Note the three Hellfire air-to-ground anti-armor missiles on the underwing pylons beneath the Predator. (Courtesy of US National Archives, Still Pictures Branch.)

Heritage Flight Forms Up. The US Air Force Heritage Flight forms up over Hampton Roads, Virginia, during its aerial demonstration, a portion of the Air Power over Hampton Roads air show at LAFB on May 21, 2004. The aircraft pictured include, from left to right, an A-10 Thunderbolt II with Capt. Matt Kouchoukos at the controls; a P-38 Lightning with Steve Hinton at the controls; an F-86 Sabre with Dale Snodgrass at the controls; and a P-51 Mustang with Ed Shipley at the controls. (Courtesy of US National Archives, Still Pictures Branch.)

Thunderbirds Aerial Demonstration. The USAF Thunderbirds Air Demonstration Team performs a breathtaking maneuver above LAFB at the Air Power over Hampton Roads air show on May 23, 2004. (Courtesy of US National Archives, Still Pictures Branch.)

27th Fighter Squadron, 1st Fighter Wing F-15Cs. F-15Cs of the 27th Fighter Squadron, 1st Fighter Wing, are pictured on the flight line at LAFB on August 12, 2004. The aircraft were taking part in a Phase 1 ORI. (Courtesy of US National Archives, Still Pictures Branch.)

Raptor Delivery. The first operational Lockheed Martin F/A-22 Raptor (foreground), flown by Lt. Col. James Hecker, was delivered to LAFB on May 12, 2005. The aircraft was the initial Raptor of a fleet of 26 eventually supplied to the 27th Fighter Squadron. (Courtesy of USAF, T.Sgt. Ben Bloker.)

Raptor AIM-120 Weapons Flight Test. In one of the first test firings of a missile from an F-22A Raptor, a 27th Fighter Squadron F-22A launches an AIM-120 radar-guided Advanced Medium-Range Air-to-Air Missile (AMRAAM), which successfully destroyed an aerial target drone above the Gulf of Mexico as part of Operation Combat Archer on February 14, 2006. (Courtesy of USAF, M.Sgt. Michael Ammons.)

Operation Noble Eagle CAP Flight. Six years after the devastating terrorist attacks of September 11, 2001, a 71st Fighter Squadron, 1st Fighter Wing F-15C based at LAFB, piloted by Capt. Matt Bruckner, performs a CAP flight above Washington as part of Operation Noble Eagle on October 7, 2007. (Courtesy of USAF, Sgt. Samuel Rogers.)

Stealth Bomber Flight. A Northrop Grumman B-2 Spirit stealth bomber carries out a flight demonstration over LAFB as part of the Air Power over Hampton Roads air show on April 25, 2009. The B-2 provides the USAF with global nuclear strike capability while remaining virtually undetectable by enemy radar. (Courtesy of USAF, A1c. Jonathan Koob.)

Trilateral Exercise Flight. From December 2 to 18, 2015, JBLE (1st Fighter Wing) served as the home base for a trilateral exercise, consisting of USAF, RAF, and French Air Force aircraft and personnel elements. Aircraft utilized in the training exercise included the USAF F-22, RAF Eurofighter Typhoon, and French Air Force Dassault Rafale. (Courtesy of USAF, Sr. Amn. Kayla Newman.)

Heritage Flight. The past and future of USAF airpower were represented in the April 24, 2016, Air Power over Hampton Roads USAF Heritage Flight at JBLE. The past was represented by a restored North American P-51D Mustang air superiority fighter from World War II, while the future was represented by the stealthy Lockheed Martin F-22 Raptor air superiority fighter and the new, stealthy Lockheed Martin F-35 Lightning II joint strike fighter. (Courtesy of USAF, Sr. Amn. Kayla Newman.)

95th Anniversary of the *Ostfriesland* Bombing. In commemoration of the 95th anniversary of the bombing of the ex-German World War I battleship *Ostfriesland*, two JBLE-based 94th Fighter Squadron F-22A Raptors drop joint direct attack munitions (JDAMs) over the Atlantic Ocean. (Courtesy of USAF, S.Sgt. J.D. Strong II.)

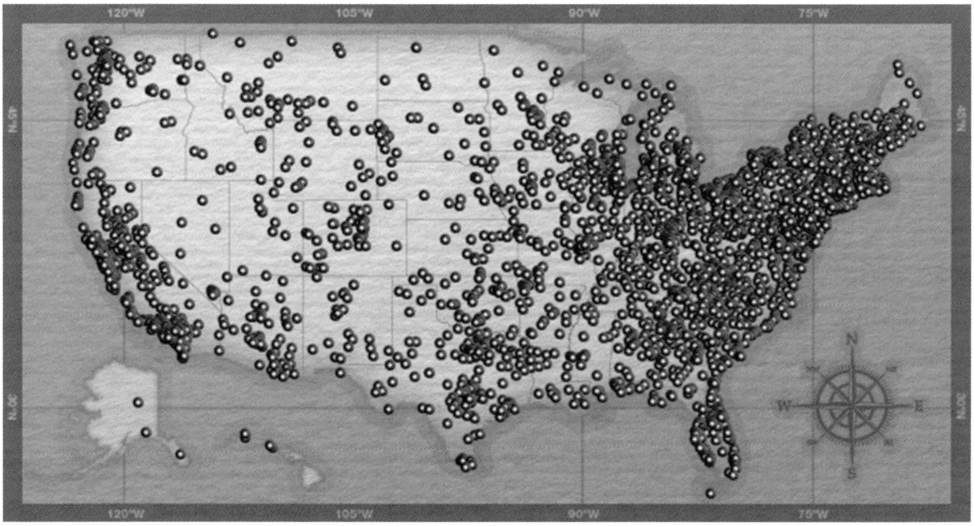